PYROTECHNIST'S COMPANION

FIREWORKS AND PYROTECHNICS
TERMINOLOGY EXPLAINED

INTRODUCTION

This publication attempts to offer the reader an understanding of the basic terminology relating to fireworks / pyrotechnics. Fireworks have been a source of entertainment and religious celebration since there first mention in 7th Century China.

Traditional fireworks produce four primary effects:

They produce:

- Light in various colors
- Noise of various intensity
- Smoke in various colors
- Floating materials such as confetti pieces or streams

The main colors used in fireworks displays are red, yellow, orange, blue, green, purple, silver and gold.

The color aspect of fireworks can be an instantaneous burst of color, comets, sparkles, or colors of various duration and intensity. Colors and effect can vary depending on their purpose of use such as cultural and religious festivals and celebrations.

Firework classifications may vary dependent upon the country and even State of use, but commonly they are heavily restricted in the West. Check your local area regulations.

The information contained within is to be used as a rough guide only and the rules, regulations and laws relating to fireworks much be researched in the area in which you operate in order to be sure you are operating within the law.

Also, remember that fireworks are can be very dangerous if not used correctly and making homemade fireworks without proper training is a recipe for disaster. Don't do it. Watch professional displays and enjoy this age old tradition.

A Rough Guide to Firework Classifications

US Department of Transportation explosives classifications

Here are some *common* fireworks classes:

- **Class 1.1G** (Mass Explosion Possible:Pyrotechnics) **UN0094** Flashpowder
- **Class 1.1G** (Mass Explosion Possible:Pyrotechnics) **UN0333** Fireworks (Salutes in bulk or in manufacture)
- **Class 1.2G** (Projection but not mass explosion:Pyrotechnics) **UN0334** Fireworks (Rarely used)
- **Class 1.3G** (Fire, Minor Blast:Pyrotechnics) **UN0335** Fireworks (Most Display Fireworks) Current federal law states that without appropriate ATF license/permit, the possession or sale of any display/professional fireworks is a felony punishable by up to 5 years in prison.
- **Class 1.4G** (Minor Explosion Hazard Confined To Package:Pyrotechnics) **UN0336** Fireworks (Consumer or Common Fireworks) Most popular consumer fireworks sold in the US.
- Class 1.4S (Minor Explosion Hazard Confined To Package: Packed As To Not Hinder Nearby Firefighters) UN0336 Fireworks (Consumer or Common Fireworks)
- Class 1.4G (Minor Explosion Hazard Confined To Package:Pyrotechnics) UN0431 ARTICLES, PYROTECHNIC for technical purposes (Proximate Pyrotechnics)
- Class 1.4S (Minor Explosion Hazard Confined To Package: Packed As To Not Hinder Nearby Firefighters) UN0432 ARTICLES, PYROTECHNIC for technical purposes (Proximate Pyrotechnics)

British fireworks classification

Main article: British firework classification

Britain has its own system of classifying fireworks.

- **Category 1** – indoor fireworks, for use in small areas.

- **Category 2** – garden fireworks; must be safely viewable from 5 meters (16 ft 5 in) and must not scatter debris beyond 3 meters (9 ft 10 in).

- **Category 3** – display fireworks; must be safely viewable from 25 meters (82 ft) and must not scatter debris beyond 50 meters (164 ft).

- **Category 4** – professional fireworks; a person must have adequate insurance and storage to purchase and use these fireworks. Insurance can only be obtained once they have acknowledged the safe use and storage of Category 4 fireworks. There is no such thing as a "license" to buy or use Category 4 fireworks.

FIREWORKS AND PYROTECHNICS EXPLAINED

GLOSSARY

1.3G Explosives	Formerly known as Class B special fireworks. Items classified as 1.3G explosives are display fireworks
1.4G Explosives	Formerly known as Class C common fireworks. Items classified as 1.4G explosives are consumer fireworks intended for use by the general public
ABS (Acrylonitrile Butadiene Styrene)	Plastic pipe used in plumbing. ABS should NEVER be used for mortars since it can shatter into razor-sharp pieces
ADR	The provisions which came into effect on 1st January 2003 concerning the international carriage of dangerous goods (including fireworks). In layman's terms the amount of fireworks you can carry in a vehicle is limited by their type, the vehicle, and whether the driver has been formally trained to transport them. The restrictions mainly apply to professional (commercial) displayers and their fireworks. See also DTR

Aerial	Any fireworks item that shoots flaming balls into the air, such as a cake or mortar
Aerial Bomb	Another term for an aerial shell
Aerial Firework	A device that functions in the air, such as a shell, roman candle, rocket, or repeater
Aerial Repeater	Same as an aerial but produces the same affect over and over
Aerial Salute	A salute that functions as an aerial shell, usually a loud explosion
Aerial Shells	A fireworks device designed to be launched into the air for use in a fireworks display, it contains a lift charge and a delay fuse for the effect
AHJ	Authority Having Jurisdiction; usually a fire marshell or anyone incharge of monitoring and regulating fireworks in your display area
Air Launching	A method of launching aerial shells using compressed air rather than a black powder lift charge. Shells are placed into a rotating turret that positions each tube

	into a firing point over an air valve. The resulting blast of air propels the shell into the sky where timed computer chips built in to the shell trigger the burst charge at the correct altitude.
Airbomb	Any shell effect launched from a firework that bangs, now banned in many areas of the world
Airbomb Barrage	Multiple airbombs fused together into one firework
Alloy	A combination of two metals that shares some of the characteristics of each. Magnalium (magnesium/aluminum), for example, is not as reactive as magnesium and not as hard to ignite as aluminum
American Pyrotechnics Association (APA)	Trade association for the fireworks industry
APA Standard 87-1	The Standard for Construction and Approval for Transportation of Fireworks, Novelties, and Theatrical Pyrotechnics

Approved	Accepted by the authority having jurisdiction
AQUA SHELL	A shell designed to be launched across, and break on, water.
Artillery	Multiple reports from a single item
Ash Can	Ash can is another name for a silver salute, true ash cans became illegal in 1966, legal ash cans contain only 50 milligrams of flash powder
Assistant	A person that works under the supervision of the pyrotechnic operator
Assortment	A variety of fireworks sold in a box, offering many different sizes, usually including fountains, spinners, rockets, and firecrackers
ATFE (ATF)	The Federal Bureau of Alcohol Tobacco Firearms and Explosives. The ATFE is responsible for regulating the sale, manufacturing, importation, storage, and use of professional display fireworks and explosives. The ATFE does not

	regulate the legal use of consumer fireworks
Atomic Pattern	A shell burst consisting of three circles on three different planes, which resembles the orbits of electrons around a nucleus
Audience	Spectators viewing a performance
Authority having jurisdiction	Any individual having the responsibility of approving equipment, an installation, or procedure. An office or an organization may also be involved, rather than a single individual
Bag Mine	A type of mine lacking a strong casing; consists of lift charge and stars within a sealed plastic bag
Ball Rocket	Popular style of rocket which mimics an aerial shell "on a stick". Generally, but not always, gives a bigger and louder effect than a standard plastic head rocket
Ball stars	Stars which burn with a spherical flame leaving no trail

Bang	A report. What most fireworks do.
Banger	A small tubular firework that simply bangs. in effect an airbomb that stayed on the ground.
Bare Match	Black match without any sort of covering or protection
Barge	An anchored water vessel from which fireworks are launched
Barrage	Rapid firing aerial fireworks, usually used in finales
BATF	Bureau of Alcohol, Tobacco and Firearms. A federal agency which oversees and regulates the safe use and handling of fireworks
Battery	A small group of similar fireworks constructed together such as missiles, fused together in such a manner that they are fired within a short period of time
Battle in the Clouds	Multiple silver sparked reports followed by a heavy report, a shell that creates several loud reports after bursting

Bee Hive	Effect that looks like swarming stars moving around
BEES	A swarm or constellate of points of light that move and disperse and dissapate under their own power. Similar to FISH, but less energetic. See Hummer
Bengal Flare	See Flare
BFA / BRITISH FIREWORKS ASSOCIATION	An association of UK firework companies who promote the safe use and sale of fireworks
Binary System	A two-component pyrotechnic system. These items are shipped as separate ingredients: an oxidizer, and a fuel. The ingredients do not become a pyrotechnic material until they are mixed
Binder	A substance used to hold certain pyrotechnic compositions together
Black Body Radiation	When light is given off by a normally dark object
Black Match	A fuse made from thread or string impregnated with black powder

Black Powder	Propellant material found in fireworks, also known as gun powder. See or blind shell
Black Shell	A shell whose time fuse fails to ignite the bursting charge and falls back to earth without bursting. Also known as a blind shell
Blinker	A small ground based firework that flashes, strobes
Blossom	A flower effect that that opens up and expands, like a flower blossoming
Blown Blind	When stars fail to ignite
Bombette	A shell effect within a cake or candle, launched by a lifting charge. Sometimes containing a variety of colors and effects
Bonfire	A fire lit in the UK on Guy Fawkes night, to celebrate a failed attempt to blow up the Houses of Parliament in London. Also known as fireworks night
Bonfire Society	Traditional English society which organises bonfires and fireworks displays

Bore	The internal diameter of a firework tube, determining the size of the shells or effects contained within
Bottle Rocket	A small rocket about the size of a standard firecracker, attached to a long stick for stabilization ending with a firecracker like report
Bottom	Fused a method of shell construction where the time fuse enters the shell at the bottom and is ignited by the lift charge
Bottom shot	The last break in a multi break shell that ends in a salute
Bottom-Fused	A method of shell construction where the time fuse enters the shell at the bottom and is ignited by the lift charge
Bounce	A black powder charge at the end of a fountain that creates a small explosion at the end of the effect
Bouquet	A number of fireworks fused together, lighting one fuse sets them all off for a concentrated effect

Bouquet Pattern	A floral shaped aerial pattern of stars, in a spherical shape
BPA	British Pyrotechnists Association. Trade body that represents professional firework display companies in the United Kingdom
Branching	Sparks that split up into smaller sparks looking like the branches of a tree
Brazilian Fireworks	Brazilian fireworks are characterized by bright colours and loud reports
Break	The burst of an aerial shell causing the effects and color or the compartment of a shell containing effects
Break of Saettine Shell	A ring of shots thrown out forcefully and uniformly thus forming a ring in the sky
Bridge Wire	A fine wire usually which heats or explodes when an electric current is applied to it. The heat causes a pyrotechnic device to ignite. Usually made from a Nickel/Chromium alloy (Nichrome), Steel, or Tinned Copper.

British Pyrotechnics Association (BPA)	Trade association for the British fireworks industry
BRITISH STANDARDS (BS) 7114	The legal standard to which fireworks sold to the public and are mainly for the benefit of user safety
Brocade	A visual effect which is like a tail effect, bright and glittery, spider like and is brighter than willow or tiger tail style bursts
Bureau of Alcohol, Tobacco and Firearms (BATF)	The federal agency that regulates explosives. This agency monitors the importation, manufacture, distribution and storage of fireworks and other explosives
Burning	An exothermic oxidation/reduction reaction. Pyrotechnics generally use oxygen rich salts such as perchlorates, chlorates, or nitrates to rapidly oxidize fuels such as metals, gums, sulphur, or charcoal.
Burst	The release of effects into the air by an aerial device

Burst Charge	A composition placed inside aerial shells which explodes at the shell's maximum altitude, breaks apart the casing and ignites the effects. Made of black powder but can also be made with potassium chlorate
Bursting Charge	The pyrotechnic composition in an shell that causes the shell to burst at a certain altitude
BUTTERFLY	An arial effect which sees two cones of effects eject in opposite directions which create a symmetrical butterfly effect
CAKE	A chain fused multi-shot firework with single or multiple effects, same as a repeated but usually larger
Caliber	The inside diameter of a mortar or the size of a shell
CANDLE	A firework consisting of a shell or effect, single or mutiple, in a card tube. A lifting charge propels the effect into the air. Commonly called a Roman Candle

Canister Shell	An arial shell shaped like a cylinder and containing any of a number of effects. Canister shells contain a larger volume of stars or inserts than round shells
Cannonade Shell	A group of shots that travel for a few instants after the shell breaks and then perform more or less together, usually with significant impact
Case	Any tube containing pyrotechnic composition
CATEGORY 1/2/3/4	The British Standards classification fireworks are given in the UK. Category 1 fireworks ("indoor") are the safest, and can be lit indoors. Category 2 fireworks ("garden") are for use outdoors and spectators must be at least 5 metres away (8 metres on fireworks labelled with EU compliance). Category 3 fireworks ("display") are for use outdoors and spectators must be at least 25 metres away, with these being the largest publicly available fireworks.

	Category 4 all other ("professional") fireworks and may only be sold to, or used by, a professional
Catherine Wheel	A circular firework that spins round and round emitting coloured fire. See Wheel.
Celebration Roll	A chain of hundreds or thousands of firecrackers traditionally used by the Chinese to frighten away bad spirits
Chain Fusing	A series of two or more aerial shells fused to fire in sequence from a single ignition
Charcoal comet	Comet star made with a charcoal composition
Charging	The process of filling a tube with pyrotechnic composition or effects
Chemical Composition	All pyrotechnic and explosive composition contained in a fireworks device. Inert materials such as clay used for plugs or organic matter are not considered part of chemical composition
Cherry Bomb	A red ball-shaped salute firecracker with high explosive power, usually illegal

Chinese Crackers	A number of small bangers strung together and connected by a rapid burning fuse, which when lit, creates a chain reaction of bangs
Chinese Lanterns	Very pretty large balloon type constructions made from flame retardant material. When a wick is lit the lantern fills with hot air and it eventually lifts off
Chlorates	Any salt of chloric acid. A powerful form of chemical oxidizer, including potassium chlorate and barium chlorate. Banned from all consumer fireworks but may be found in some professional fireworks
Chlorine Donor	A chlorine-rich compound such as PVC (polyvinylchloride) or Parlon. When combined with a metal within a pyrotechnic flame, certain colors can be produced.
Choke	The narrow portion of a rocket or fountain tube through which the effect or proppelant is forced which therefore

	increases the velocity of the products being ejected or to create thrust
Chrysanthemum	A dense, spherical burst of stars that retains its shape before fading where tailed stars are thrown out from the center creating a burst of spokes and an expanding globe of color. This is the most well known type of firework shell break
Class B	Obsolete DOT classification for 1.3G display fireworks, though still commonly used amongst those in the fireworks business
Class C	Obsolete DOT classification for 1.4G consumer fireworks, though still commonly used amongst those in the fireworks business
Coconut Tree	Similar to a Palm Tree shell
Cold Fall Out	Fall out that is not burning such as indoor ice fountains have cold fall out
Color	A single burst of color stars

Color & Report	A single asymmetrical break of colored stars followed by a heavy report - bang
Color Changing Shell	Effects that change from one color to another
Color Changing Spider	Hard break of gold tailed stars that change to a bright color at termination
Color Pot	Tube which is filled with pyrotechnic material that produces a colored flame
Color, Whistle & Report	An asymmetrical break of colored stars with whistling or screaming followed by a loud report
Colored Smoke	A chemical composition which produces dense clouds of colorful smoke when ignited
Comet	A large cylindrical pumped star with a pellet composition which is propelled from a mortar or shell and produces a sparked tail, much like aerial display shells
Commercial Fireworks	Fireworks professionals prefer to use the term "Display Fireworks" for the larger fireworks that are used in public

	displays, and usually avoid the term "Commercial Fireworks"
Composition	A mixture of pyrotechnic chemicals which contains a fuel, an oxidizer, and sometimes various colur producing chemicals
Concussion effect	An effect that produces a very loud and jarring shock for a dramatic effect
Concussion mortar	A device specifically designed and constructed to produce, and contain a loud noise and a violent jarring shock for dramatic effect. Aslo known as a "Maroon", "Cacoo" or "Sonic Boom pot"
Cone	A cone shaped fountain
Confetti	Small strips of colored streamers used mostly in novelties or at indoor fireworks displays which are propelled by a gas cartridge or by a small pyrotechnic charge
Confetti Cannon	A tube that fires confetti, streamers or other materials which are propelled by a

	gas cartridge or by a small pyrotechnic charge
Conic Fountains	A type of fountain. See Fountain
Consumer Fireworks	Also known as 1.4G fireworks. Fireworks that are intended for use by the consumer
Consumer Product Safety Commission (CPSC)	Federal agency which regulates consumer 1.4G fireworks.
Continuity test	A test to find whether an electrical circuit works
Convolute (Parallel) Tube	Paper firework tube wound in a parallel fashion which is stronger than spiral wound tubes
CPSC	The US Consumer Product Safety Commission, a federal agency responsible for testing and approving all consumer fireworks
Cracker	A nickname for a firecracker that has an audible effect that sounds like a crack
Crackle	Clusters of small, crackling, snaps, pops, and flashes and sharp reports usually

	accompanied by an aerial gold lace visual effect
Crackling Comet	A comet that leaves behind a tail of crackling effects rather than just glittering colour
Crosette	A star that burns for a period of time then explodes from an internal shot with smaller fragments flying outward from the star's trajectory with all stars in one timing usually breaking simultaneously. The Fragments fly in different directions crossing over themselves. Also known as a Splitting Comet
Crossmatch	A technique used to ignite the time fuse in shells by having a piece of black match threaded through a hole in the time fuse, so fire is transferred from the black match to the black powder core of the time fuse
Crown	A Chrysanthemum with the addition of a inner pistil or petals of color stars and outer stars with bright, short duration

	color core. Also known as a Diadem Chrysanthemum
Cut Stars	Cubical stars cut from damp pyrotechnic composition with a knife
Dahlia	Symmetrical burst pattern with a long duration creating a dropping effect, similar to a peony, but with larger and fewer stars
Dahlia shell	A shell with relatively few comets, often producing a starfish like shape
Damp Squib	A firework that fails to ignite or explode
Dark Fire	A composition that emits almost no light as it burns. The star will burn one colour, "burn out", then appear to ignite again in a different colour. Also known as a dark prime
Day Box	A portable magazine that is used for temporary storage of pyrotechnic materials
Day Time Effect	Any firework that can be enjoyed during day time such as smoke and parachute items

Daylight Shell	A shell designed to be fired during the day containing effects such as smoke, reports and whistles
Deflagrate	Cause to burn rapidly and with great intensity or to burn or vaporize suddenly or to burn with great heat and intense light, usually accompanied by a considerable amount of heat and large volumes of gas. When the speed of the burn or escaping gas exceeds the speed of sound, the result is a loud boom. Deflagration is the scientific term for how fireworks explode
Deflagration	A rapid decomposition reaction which is accompanied by the evolution of light, heat, and large volumes of heated gas
Delay	A pyrotechnic composition that is used for the timings between ignitions such as in a roman candle
Department of Transportation (DOT)	Federal agency which controls the transport of all hazardous materials including fireworks

Detonate	Cause to burst with a violent release of energy. Burst and release energy as through a violent chemical or physical reaction. A characteristic of high explosives, a detonation occurs when the explosive decomposition of a substance forms an energy wave that moves rapidly though the substance at speeds that exceed the speed of sound
Detonation	A violent release of energy caused by a chemical or nuclear reaction causing a shockwave producing explosion
Detonator	A small explosive used to set off high explosives and not to be confused with firework electric igniters
Discharge Site	The area immediately surrounding the fireworks mortars
Display Firework	A large professional firework designed to produce visible or audible effects for entertainment purposes and requiring 25 metres distance to spectators

.

Display Permit	A special permit that is granted by the local authority to allow you to shoot fireworks legally
Display Site	The immediate area where a fireworks display is conducted
Divisional Storage	The name given to a type of professional fireworks storage used to store large amounts of product
DIY KIT	A kit made up of numerous items to make a complete display
DOT	Department of Transportation the Federal Agency responsible for regulating the labeling and transportation of fireworks in the United States
Double Break	A firework, rocket or shell that has two different effects
Dragon Eggs	Clusters of strobing, crackling, popping sparks in the air usually in silver or gold
Draw Out	A color break followed by 4 timed shots, followed by another color break and ending in a final report

Drivers	Thrust producing fountains used to propel devices such as wheels
Drizzle	Falling effect of glitter that resembles rain or showers
Dross	Hot, molten waste product of combustion
DTR	In relation to ADR, DTR refers to the training required by drivers of vehicles transporting dangerous goods including fireworks. ADR specifies limits of fireworks above which driver training is required.
Dud	A firework that did not light or produce their desired effect
EIG / Explosives Industry Group	A UK organisation that "exists to represent and inform its members on all topics of explosive legislation in the UK"
Ejects Bangs and Ejects Stars	Common description on firework labels. EJECTS STARS it is likely to be fairly quiet, whereas EJECTS BANGS is likely to be noisier

Electric Comet	Comet star where the sparks in the glowing trail give the impression of electrical sparks
Electric Firing	Igniting fireworks with an electric charge
Electric Igniter (electric matches)	Device used for the electrical ignition of fireworks which consists of two lead wires connected to each other by a small filament of nickel chromium (nichrome) wire coated with pyrogen. When a current is passed through the igniter the nichrome filament heats up and ignites the firework. Often called Squibs (incorrectly)
Electric match	A device containing a small amount of pyrotechnic material that ignites when a specified amount of electric current flows through the leads
Electrical Firing Unit	A panel or box with manually operated switches that control the electric current used to ignite fireworks
Electrical Ignition	A technique used to ignite fireworks using a source of electric current

Ember	A burning piece of casing or paper from a firework
EX Number	The identification number assigned by DOT to a commercial fireworks device
Exothermic	A chemical reaction or compound occurring or formed with the liberation of heat. A chemical reaction in which the total energy of the products is less than the total energy of the reactants. Firework reactions are exothermic
Explosive	A chemical substance that undergoes a rapid chemical change (with the production of gas) on being heated or struck. Explosives fall into 2 classes, detonating and deflagerating
Explosive Composition	Any chemical compound or mixture, the primary purpose of which is to function by explosion
Fall Out	Debris that falls to earth after a firework has ceased to produce effect, usually just card casing and paper

Falling Leaves	A beautiful aerial effect where sparks create drooping pattern from a central point
Fallout area	The area in which any hazardous debris falls
Fan Cake	Usually a major display cake where the tubes are angled, sending shots left and right of the display area in a fan
Finale	String of shells fused together to fire rapidly in order from a rack of tubes
Fire	To ignite pyrotechnics
Fire Cracker	A firework producing a loud report
Fire Writing	See Lancework
Firefly	Also known as Transformation. A gold or silver flashing effect occurring in the tail of a star where the effect is sustained and appears to hang in the sky
Firework	An item consisting of pyrotechnic composition producing audible and visual effects
FIREWORK CODE	Issued by the DTI, a "layman's" guide to firework safety

Fireworks Display	A presentation of fireworks for a public gathering
Firing current	The amount of current required to ignite an electrical igniter
Firing system	The source of ignition for pyrotechnics
Firing Technician	Individual who ignites fireworks devices at a show
Fish	A type of aerial effect that is similar to the bee hive effect except the stars swarm outward faster. The effect is created using small chunks of fast burning fuse that actually propel themselves
Fish & Whistle	The fish are silver tailed serpents, the whistle is multiple screaming serpents
Fixed Production	Any production performed repeatedly in only one geographic location such as a tourist attraction or site
Flame Projector	A device shaped liked a mortar tube that creates a short-lived fireball or flame effect

Flammable	Easily ignited, as are most ingredients used in fireworks
Flare	A pyrotechnic device that is designed to produce a single source of intense light
Flash Pot	A device used with flash powder that produces a flash of light, smoke, sparkles or an audible report
Flash Powder	Pyrotechnic mixtures which contain powdered aluminum or a magnesium/aluminum alloy which, when ignited, can result in a violent explosion and flash.
Flitter	A bright spark trail effect that does not split into sparks, left behind by a star
Floral Pattern	An aerial pattern that resembles a flower
Flowerpot	When a shell explodes prematurely in the mortar, spraying the effects into the air like a mine
Flying Saucer	A device made from gerbs or motors mounted in a circular fashion which create lift and spin
Fountain	Firework that emits showers of sparks

Fuel	Combustables or chemical-reducers such as: sulfur; aluminum powder; iron powder; charcoal; magnesium powder; magnalium powder
Fuse Cover	The protective safety cover on fuses
Fusilading Shot	A series or group of shots that travel after the shell breaks and performs randomly over a short period of time
Gabe Mort	A sack of flash powder suspended from a frame which is ignited to create a deafening blast and concussion
Garden Firework	A firework usually requiring 5 metres distance to spectators. In the U. K., a small consumer firework designed to be used in small, confined outdoor areas
Garden Pack	Normally very small and are not suitable for larger displays
Gerb	A cylinderical preload device typically made from a heavy walled cardboard that is intended to produce a spray of sparks, sometimes used with lance-work set pieces

Girandola	A wheel shape that spins horizontal flying in the air, emitting a spray of sparks and, sometimes, a whistle then explodes
Glitter	Composition producing sparks that are ejected from the burning star, consists of bright flashes of light and small explosive bursts
Go Getter	A self propelled star that "swims" in the sky link a rocket without fins
Goggles	Essential eye protection for firers.
Green Man	The name for pyrotechnicians from the 15th century who wore green leaves and mud as both a protection and camouflage from the crowd
Green Mix	Essentially a raw mixture of black powder ingredients that haven't been properly combined with heat to create real black powderpolverone or pulverone. Also known as Green Powder
Ground Display Piece	A pyrotechnic device that functions on the ground

Ground Firework	A consumer firework that functions at ground level, such as fountains and smoke items and usually does not shoot objects into the sky
Ground Salute	A salute that functions from a stationary, secured position
Ground Spinner	An item that spins on the ground producing sparks
Gun	Nickname for a mortar
Gun Powder	A black coloured powder comprising in its basic form three ingredients sulphur, charcoal and potassium nitrate. See Black Powder
Guy Fawkes	The man who tried to blow up the Houses Of Parliament in the Gunpowder Plot which is celebrated in the UK on November 5th each year
Hammer Shell	A color break followed quickly by a report, another color break and report
Hangfire	A fuse or pyrotechnic ignition composition which continues to burn slowly instead of burning at its normal

	speed and may suddenly resume burning at its normal rate after a delay. If the hangfire is extinguished it is termed a misfire
Hazardous Debris	Any debris that is capable of causing personal injury or property damage
HDPE	High Density Polyethylene. A strong plastic pipe used for mortars
HEART	An effect that creates a heart shape
Heart Pattern	Colored stars that form a heart pattern in the sky
Helicopter	A device that spins into the air producing an aerial effect
High explosive	A very powerful explosive, such as TNT or dynamite, not used to make consumer fireworks
High Level Fireworks	Devices propelled into the air, usually aerial shells
Holder	Any device used to hold a pyrotechnic device or material other than a Mortar
HUMMER	A firework shell or projectile that makes a "humming" noise

Hygroscopic	The property of a chemical composition that causes it to absorb and retain moisture from the air
Ice Fountain	A fountain with Cold Fall Out and low smoke, normally designed for indoor or stage use.
Igniter	An electrical, chemical, or mechanical device normally used to fire pyrotechnics
Indoor Firework	A small firework which can be safely lit indoors
Ingeredient	A chemical used to create a pyrotechnic material
Initiator	A device containing primary explosives that is used to initiate quantities of high explosives. These are not fireworks.
Instantaneous Fuse	Black match that is encased in a loose-fitting paper or plastic sheath to make it burn extremely rapidly. Quickmatch is used for simultaneous ignition of a number of pyrotechnic devices such as lances. Also known as Quickmatch.

Integral Mortar	A preloaded mortar containing pyrotechnic materials and intended for a single use only
Jeweled Rats	A rocket that flies in a straight line that is also carrying effects on the outside such as stars
Jumping Jack	Small fire crackers that appear to jump off the ground
KAMURO	An effect that hangs and trails in the air not unlike a willow, often strobingand used effectively in finale sequences to fill the sky
Kamuro Chrysanthemum	A single-petaled chrysanthemum break with tailed stars of significant duration to create a pronounced and uniform willow effect
Labels	All legal consumer explosives have mandatory labeling requirements
Lady Finger	A very small, thin firecracker
Lance	A small tube of pyrotechnic composition that burns with a steady, flare-like flame

	for approximately one minute in various colours
Lancework	Set piece such as a sign that lights up with numerous pyrotechnical devices and can spell out "Happy Birthday" for instance
Leader	The fuse that transfers fire from the electrical igniter or day fuse to the lift charge
Lift Charge	The black posder charge that propels as shell into the air
Loader	An assistant who loads or reloads aerial shells, comets, or mines into mortars
Loose items	Fireworks that are not part of a larger pack or kit
Low Explosives	Explosives that burn at a steady speed and can only be detonated under extreme circumstances. Examples of low explosives are black powder and fireworks.
Low Level Fireworks	Fireworks that ignite and fire their effect into the sky directly from the ground.

	Any of a class of fireworks devices that either perform below approximately 200 feet (60 m) or begin their display at ground level and rise to complete their effect
Low Noise Fireworks	A firework specifically designed to operate with little or no noise
M 80	A small, powerful explosive created by the military for use as a grenade and gunfire simulator
Magazine	Any building or structure, used exclusively for the storage of explosive materials
Magnalium	A mixture of aluminium and magnesium
Manual Ignition	A technique used to ignite fireworks using a handheld ignition source such as a portfire
Manufacturer	An individual who prepare pyrotechnic material or is involved in there loading or assembly
MAROON	A very loud bang typically created by a maroon shell. British term for a salute

Matching	The process of connecting multiple fireworks with quick match
Mine	A device containing pyrotechnic effects which are simultaneously ignited and dispersed into the air from mortar
Mine (star mine)	A firework similar to a shell that ignites effects such as stars, launching them into an aerial display
Mini-Rocket	A small screech type rocket
MISFIRE	A firework that goes off incorrectly
Missile	Tube with fins on it that shoots into the sky producing report and sometimes whistles
MODE A / MODE B	Under the old regulations which have now been superceded this referred to the most basic registered firework storage for personal or retail use. Mode B allowed the storage of up to 250Kg of shop good fireworks, Mode A allowed up to 1000Kg. Replaced now by MSER. More info.

Monitor	The individual at a fireworks display responsible for observing the perimeter of the firing site and insuring that security personnel or barriers keep spectators at a safe distance
Morning Glory	Large sparkler that changes effects and colors
Mortar	Paper or HDPE tube for firing shells and effects into the air, some tubes can be reloaded with shells whilst others are single use only
Mortar Rack	Rack which contains multiple mortar tubes
Mortar Trough	Above ground structure filled with sand or similar material into which mortars are positioned
Motor	The part of a rocket that burns to give the rocket lift
MSER	Manufacture and Storage of Explosives Regulations

Multi	break shell with numerous compartments, each one bursting separately.
Multi break shell	An aerial shell comprising more than one section producing a separate effect in sequence and ignited by the bursting of the preceding section
Multi Shot Aerial	Another name for a cake or repeater
MULTI-SHOT	A firework that has more than one shot.
Muzzle Break	Term for a shell exploding right after leaving the tube or mortar sending the effects all over the ground
National Fire Protection Association (NFPA)	Organization which provides several standards that outline recommendations for the manufacture, storage, transportation, and execution of fireworks
Nautic	Firework launched into water and then floats on surface
NFPA	National Fire Protection Association. A national agency involved with enhancing

	and regulating the construction, handling, and use of fireworks
NFPA Standard 1123	Code for Fireworks Display
NFPA Standard 1124	Code for the Manufacture, Transportation, and Storage of Fireworks and Pyrotechnic Articles
NFPA Standard 1126	Standard for the Use of Pyrotechnics Before a Proximate Audience
Nosing paper	Thin paper wrapped around the nozzle of a pyrotechnic device to hold the fuse in place and prevent premature igniting of the device
Nova	A break of red and blue stars followed by a ring of silver sparked reports
Novelties	Small fireworks such as party poppers, snakes and sparklers
Occupational Safety and Health Administration (OSHA)	Federal agency that inspects fireworks manufacturing plants. OSHA not only regulates non-fireworks specific aspects of plant safety (i.e. housekeeping, electrical requirements, employee

	training), but also the fireworks-related standards of NFPA Standard 1124.
Old Spec / Stock	A term to describe their consumer fireworks that pre-date new fireworks regulations, usually higher powered
Operator	The person with overall responsibility for the operation and safety of a fireworks display
Operator Fired Display	Term used to denote a professional display
Orange Book	Nickname for the booklet titled "ATF Explosives Law and Regulations"
Oxidizer	A substance that oxidizes another substance. Usually an oxygen-rich, ionically bonded chemical that decomposes at moderate to high temps
Palm Tree	A comet shell whose rising tail is the trunk while the palms are made by large stars leaving a trail in a palm pattern, "branches" of sparks

Palm-tree shell	A shell with usually charcoal or gold comets which bursts to give the impression of palm fronds
Parachute	A projectile that explodes into a parachute that gently flows towards the ground, some can have a strobe effect, smoke or sparks following them
Parallel burning	A sequence where a piece of burning material ignites the piece next to it, which in turn ignites the piece next to that
Parallel matching	Ignition sequence where one fuse is connected to and simultaneously ignites multiple pyrotechnics devices
Party Popper	A small canister consisting of confetti that shoots out when u pull the string
Pattern shell	A shell that breaks in a perfect spherical pattern
Pearl	A single colour star, launched from the ground
Peony	A shell of ball stars. A loosely symmetrical break of stars without trails

	that fly outward and then begin to droop downward
Perchlorates	A common oxidizer used in fireworks manufacture, perchlorates are preferred over chlorates because their compounds are generally less sensitive to shock
Persistence	The amount of time an effect is visible before it disappears
PGI	Pyrotechnics Guild International
PIC	A yellow, waterproof fuse used by professional displayers
Pigeon	A device that consists of many rats, designed to fly back and forth
Pin Wheel	A type of Catherine wheel. See Wheel
Pipe	Loose paper tubing fitted over black match to make quick match
Pistil	Where an effect such as a peony has a central part to it. The central part is known as the pistil or A ball of stars in the center of another ball of stars

Planes	A term for a device that spins very fast and lifts into the sky, then explode or burst into a special aerial effect
Port fire	A long tube containing slow-burning pyrotechnic composition that is used to ignite fireworks
Prime	A composition such as black powder that is relatively easy to ignite that is used to help ignite something that is more difficult to ignite
Professional (1.3G) Fireworks	See Definition And Types Of Professional Fireworks
Professional Firework	A firework with an unclassified safety distance for use by professionals only
Proximate Audience	An audience that is closer to pyrotechnic devices than allowed or a formal name for indoor fireworks
Pulverone	See Green Mix.
Pumped Stars	Stars produced by compressing star composition out of a cylindrical tube like a syringe, and cutting them to a specific length

Punk	Slow burning wood used to light fireworks usually made from compressed sawdust
Pupatelle /Pupadelle	Small insert shells that are inserted to produce a splash of color
PVC	A type of plastic pipe, polyvinyl chloride, which is dangerous if used to make for mortars due to its tendency to break into shrapnel if a firework flowerpots
Pyro	From the Greek word for "fire"
PYROMANIAC	Someone with an obsessive desire to set things on fire, but also used to describe someone who loves fireworks
Pyrotechnic Composition	A chemical mixture which on burning and without explosion produces visible displays of lights or whistles etc
Pyrotechnic Device	Any device containing pyrotechnic materials and capable of producing a special effect
Pyrotechnic Material	A chemical mixture to produce visible or audible effects by combustion, deflagration, or detonation

Pyrotechnic Special Effect	A special effect created through the use of pyrotechnic materials and devices
Pyrotechnician	One who makes and / or shoots fireworks, and is Licensed to handle and use pyrotechnics
PYROTECHNICS	A general term meaning the art of making or displaying fireworks
Pyrotechnist	Someone who is proficient in shooting fireworks displays
Quarter Stick	Quarter sticks were similar to M80's, but were larger in size and contained more of flash powder
Quick match	A fuse that burns extremely quickly. Also known as an Instantaneous Fuse
Rack	A frame used to hold mortars or launch rockets
Rainbow	Used to describe a firework's effect when it changes colour. Numerous colours from the same product such as red, green, yellow and orange in succession

Ramming Rod	A rod made usually of wood which is used to compress pyrotechnic compositions within a tube
Rat	A rocket constrained to fly along a line and emitting effects such as sparks
Ready Box	A storage container for aerial devices for use during set-up and display
Ready Box Tender	Assistant who controls and dispenses the contents of ready boxes
Reloadable Aerial	An aerial mortar that includes one or more mortar tubes and several reloadable aerial shells. The shells are placed inside the mortar tube, a long quick burning fuse is lit, and the item is fired into the sky. These items are consumer versions of the mortar based fireworks used in professional display fireworks.
Repeater	A firework which shoots multiple aerial effects into the air
Report	A salute or a loud concussion, which explodes violently

Report Salute	A very loud bang typically created by a maroon shell
Retire Immediately	Warning on firework labels
Ring Effect	A two dimensional, expanding ring created with multiple stars and seen in professional displays
Ring Shell	A shell that produces a ring as its aerial pattern
Rising effect	A tail or whistle that starts as it leaves it base
Rising Tail	Shells that leave a tail from the mortar up until they break
Rocket	A firework that is propelled into the air by a fast burning engine before releasing its effects
ROCKET POD/VOLLEY	A box or tube containing multiple small rockets fused together creating a rocket salvo which is very effective
Roman Candle	A chain-fused firework that propels a series of aerial shells, comet or mine effects into the air from a single tube

Saettine	Shot originally made with dark flash, thus producing a report with low light output
Saftey Area	The space between fireworks and spectators as set by law
Safety Cap	A tube, closed at one end that is placed over the end of the fuse until intended ignition to protect it from damage and accidental ignition
Safety Fuse	Used to delay the time of firing for the operator to move away, also known as visco
Saltpeter	Older term for potassium nitrate (KNO3) the most common type of oxider used in fireworks
Salute	A brilliant flash or light followed by a report
Salute Powder	A pyrotechnic composition that makes an explosive sound when ignited
Salvo	Rapid simultaneous discharge. A rapid sequence of effects or fireworks
Saturation	The concentration of a firework's effect

Saturn Ring	A peony with a colored ring surrounding it which resembles the ring around the planet Saturn
Screech Rocket	A smaller rocket whose main effect is to screech loudly in flight and which usually ends with a loud salute
Selection Box	A box containing a number of fireworks with a broad range of effects
Sequential multi-break	Shell which bursts into smaller shells which burst in a timed sequence
Series Circuit	An effective way of connecting electrical igniters, allowing multiple shots to be triggered from a single circuit
Series Matching	ignition sequence where devices are fused together using one or more types of pyrotechnic fuse allowing them to be ignited one after another
Serpent	Another name for a tourbillion. A small tube producing a visual tail tracing an often erratic course

Set Piece	A ground display such as lancework, wheels, gerbs or fountains, that are attached to the ground
SFX	Special Effects. Audible or visual effect used to create an illusion in stage, theatre or film
Shell	A short term for Aerial Shell
Shell of Shells	A ball or cylinder filled with varying compositions. This shell contains smaller shells inside. These shells are thrown out into a wide circular pattern producing a ring of small bursts of effects
Shooter	An individual that ignites fireworks at a show
Shop Good Fireworks	Category two and three fireworks for sale to the general public
Short circuit	Ignition failure
Shot	The number of effects in a fireworks device
Side Spit	Sparks and flame that shoot out from a fuse as it burns

Silver Salute	Illegal explosive similar to an M 80 with a silver tube
Single Shot Aerial	A single shot aerial is a mortar tube with a shell already installed in it
Site Assessment	A pre-setup check of a proposed site checking its suitablity for a fireworks display, has adequate safety distances and satisfies relevant health and safety concerns
Site Plan	A sketch or map of the fireworks display site
Sky Flyer	A device that spins very fast and lifts high into the sky bursting into a special effect. Also known as planes, helicopters, or UFOs
Sky Rocket	A pyrotechnic device made out of a paper tube and attached to a long wooden stick that propels itself into the air in order to fly
Smoke	A dispersion of fine solid particles in air
Smoke Ball	A small ball that produces different colors of smoke

Smoke Item	A firework that generates smoke as its primary effect
Smoke Pot	A device that is either preloaded or loaded just before use to produce colored smoke
Snap Dragon	Asymmetrical break of variegated stars with screeching whistles
Snaps	Small pellets that ignite and produce a salute when thrown against a hard surface
Spark	Light and heat emitting particle ejected from a burning composition
Sparkler	Small coated wire that produces sparks when lit
Spindle	A spike
Spinner	A type of star that spins in the sky and gives off large quantities of white light. Also known as tourbillion
Spiral-wound tube	A type of tube created by winding multiple strips of thin paper at an angle
Spotter	A member of the fireworks display crew who observes the firing and bursting of

	aerial shells and other display items and watches for defects
Squib	A type of electric match used to set off explosives
Star	A pellet that produces an effect
Star Gun	A small roman candle like device used for testing stars
Star in a Ring Pattern	The colored stars form a star pattern in the sky. Around this star is a ring pattern of colored stars
Star Pattern	Colored stars which form a star pattern in the sky
Star Pump	Syringe like container through which star composition is pushed out of and cut into individual stars
Stars	Small balls shaped masses of mixed and hardened pyrotechnic compositions that are projected from devices such as aerial shells, roman candles, and mines
Stickless Rocket	Rising gold or silver fireball that leaves a long sparked tail

Sticky match	A fuse consisting of a trail of black power between two pieces of tape stuck against each other
Strobe	Bright stars that each flash repeatedly
Strobe pot	Ground-based firework
Tadpole	A bright wiggling worm effect with a tail
Tail	A burning tail of sparkles that follow comets or willows with a glitter like effect
Theatrical Pyrotechnics	Pyrotechnic devices for professional use in the entertainment industry
Three Break	Three timed breaks of different colors followed by a heavy salute / report
Thunderflash	A pyrotechnic device that explodes with a loud bang accompanied by a flash and used in battle simulations
Tiger Tail	A rising fireball that leaves a soft glittering tail which then breaks into an effect
Time fuse	Slow burning fuse used to create time delays in aerial shells

Titanium Report	A loud explosion in the air with white sparks
Top	Fused a method of shell construction where the time fuse enters the shell at the top and is ignited by the leader fuse
Touch Paper	Paper that burns slowly but steadily
Tourbillion	Tube of composition with a hole or holes causing a spiral like effect as the tubes are propelled through the air
Tube	Another name for a mortar, holds aerial shells
UFO	A device that spins very fast and lifts high into the sky then explodes into an aerial effect
V-FIRING CAKE	A display cake that fires two columns of effects to the left and right of the display area in the shape of a "V"
Visco	Safety fuse
Volley	An intense barrage of shells or rockets
Wall of Fire	A set of 12 or more tailed shells connected electronically to fire simultaneously giving the effects of a

	rising wall of gold or silver sparked comets
Water (Aquatic) Shell	A shell designed to function on the surface of water
Water Fall	These are held in place upside down in rows of several pieces from stage trussing creating a downward shower or water fall of sparks
Weeping-willow shell	A shell which contains very long buring charcoal comets of a long duration which fall almost to the ground giving the impression of a weeping-willow tree
W-FIRING CAKE	A display cake that fires three columns of effects – one vertical and one either side, to create a three-spoked or "W" shape
Wheel	A round device that spins rapidly creating a circular effect of flames or sparks
Whirlwind	A tube that spins in the air emitting showers of sparks

Whistle	A small tube filled with a fierce burning composition causing a whistling effect as it burns
Whistle Mix	A composition that uses potassium/sodium benzoate as a fuel and creates vibrational burning making a whistling sound
Whistles	Products producing a loud shriek sound
Whizzer	A small rotating device fired from candles, mines or shells which rotates so fast it makes a humming or whizzing sound as it flies
Wholesale Fireworks	Fireworks that are sold by the case
Willow	An aerial effect of slow falling trails of sparks
X CAKE	Cross-firing fan cake
Z-FIRING FAN CAKE	A fan cake that fires its shots in sequence from left to right or vice versa, rather than in complete simultaneous waves.